Blocks, Bears and Building Math Skills

Jeanne James
Pat Biggar

Cover Design & Illustrations by **Craig Rogers**

Icons by **Tim Bailey**

Dedication

To Anne Sanford: her wonderful ideas and dedication have impacted the young children of the world.

To our mothers; Jean Cunningham and Peg Doughty: their early support enriched our love of learning.

To our children; Brett, Jason and Allison: their joy in mathematical ideas and activities is limitless.

To our husbands; Lance and Lee: their support has enabled us to share our ideas with many young children.

Blocks, Bears and Building Math Skills
Table of Contents

Activities for Five-Year-Olds **57**

Introduction

Blocks, Bears and Building Math Skills is designed to help very young children develop the skills they will need to succeed in math. It draws on the child development theories of Jean Piaget and the recommendations of author/educator Sheila Tobias.

What Can We Teach Young Children About Math?

In her book, *Overcoming Math Anxiety*,[1] Sheila Tobias recommended that educators give children opportunities to develop their intuitive strengths. She encourages math educators to concentrate on the *process* of mathematics, rather than on the "right answer." Activities which involve estimating, guessing and deciding which might be the right answer can help children develop their own problem-solving strategies. Tobias advocated structured exploration of concrete materials to help children develop mathematical thinking.

According to Jean Piaget, recognizing similarities and differences and sequencing by size are critical pre-math skills.[2] Appropriate activities for young children include matching, sorting and categorizing. Children can classify objects (blocks, sticks, buttons, pebbles), discriminate object qualities (sizes, shapes textures, colors) and identify relationships (up and down, far and near, behind and in front of). These activities will prepare them to master the more complex math skills they will be expected to learn in elementary school.

The activities in this book are based on the theories of Piaget and the recommendations of Tobias. They are organized into five skill categories shown here with the icons used in the book:

Shapes - Pre-geometry and pre-fractions skills which involve shape recognition and the identification of parts and wholes;

Classification - Pre-numbers and pre-measurement skills which include recognizing similarities (matching, sorting) and categorizing by attributes (color, shape, weight);

Ordering - Pre-numbers and pre-measurement skills which include recognizing differences and ordering by size (linear: tall and short; liquid: empty, half-full, full);

Sets - Pre-math operations skills which include recognizing more and less and comparing sets for likenesses and differences;

Counting - Pre-math operational skills (e.g. one-to-one correspondence), identifying and understanding the values of coins or money; and,

Other Fun - Math-related activities including pre-probability and statistics skills (making and checking simple predictions and making graphs).

How to Use This Book

After the introductory material and the Math Skills Ladder, the book is divided into three sections, by age. (See Figure 1.) For each activity in the three-year old section, there are parallel activities in the four- and five-year old sections. There are, for example, three related activities about shapes:

> For three-year olds - "Shape Match Grab Bag;"
> For four-year olds - "Shape Match Treasure Hunt;" and,
> For five-year olds - "Shape Match Bingo."

With three levels of difficulty for each activity, you can provide individualized instruction for groups which have children of different ages or skill levels.[3] The titles of related activities are similar (as in the above titles). You may, of course, substitute other available manipulable materials for those suggested in each activity.

Book Organization

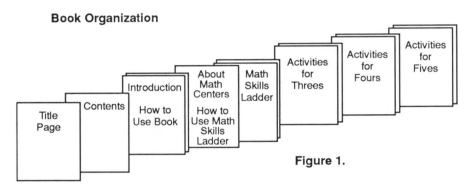

Figure 1.

Each section contains activities for the five skill categories mentioned above. Each skill category is identified by an icon in the upper right-hand corner of the page. (The icons are shown next to the explanation of each skill category on the previous page.)

Each activity appears on a single page which includes the icon of the skill category, the title of the activity, a description of skills the children will learn from the activity, and materials which are needed. Also included is information about what you need to do ahead of time (either in preparing materials or setting up the activity for the children), activities which the children do and, finally, additional activities related to the materials or skills. (See Figure 2 for the layout).

Activity Page Layout

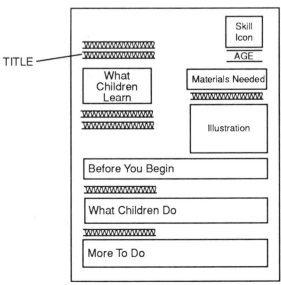

Figure 2.

About Math Centers

You may want to set up a "Math Center" in your classroom. There, you can provide a wide variety of materials which will help children develop math-related knowledge and skills through hands-on exploration and free play. Here are suggested materials to place in your math center:

Grocer's scale	Balance scale	Straws
Blocks	Counters (teddy bears, chips, pennies)	Toothpicks
Measuring spoons	Cup measures	Balls
Dice	Dominoes	Wooden Beads
Craft sticks	Pattern blocks	Bottle Caps
Unifix Cubes	Plastic soda bottles (varying sizes)	Play Money
Rulers	Items to measure	Pitchers
Pegboard	Funnel	Toy Cash Register

How to Use the "Math Skills Ladder"

Young children learn new skills in a particular sequence and the learning process itself is sequential in nature. Basic math concepts are, therefore, organized into a hierarchy by asking Robert Gagne's[4] question, "what does the learner have to already know how to do so the next task can be learned?" Children start by working on the lowest-level skill (the bottom rung on the "ladder" in each skill category) and move progressively upward as each new skill is mastered. Some skills may, of course, develop simultaneously (e.g. discriminating between heavy and light, large and small).

Remember, young children should be encouraged to learn at their own pace. What is important is that each child master one skill before moving on to the next, not that all children move through the sequences at exactly the same rate. It may take a year or more for many of these skills to develop. Teachers are encouraged to provide plenty of time for free exploration and discovery, as well as time for practice with the more structured activities provided in this collection.

Math Skills Ladder

❏ Classifying objects by two properties (color, size, shape, texture).
❏ Classifying objects by one property (color, size, shape, texture).
❏ Grouping objects by one property (e.g. size, shape).
❏ Sorting objects by one property (e.g. size, shape).
❏ Discriminating between loud and soft.
❏ Classifying objects by property (color, size, shape)

Classification Skills Checklist (Identifying Similarities)

❏ Arranging parts as below and verbalizing the reason for the arrangement.
❏ Ordering/arranging objects/shapes by size: big, bigger, biggest; small, smaller, smallest, long, longer, longest.
❏ Identifying long/short, tall/short objects.
❏ Stacking rings according to size.
❏ Identifying full/empty.
❏ Discriminating between heavy and light-weight objects.
❏ Identifying large and small objects.

Ordering Skills Checklist

❏ Identifying halves and wholes.
❏ Matching and identifying shapes listed below plus rectangles and diamonds.
❏ Matching and identifying shapes (circles, squares, triangles).
❏ Discriminating between circles and squares.

Shapes Skills Checklist

❏ Establishing equality in 2 sets by making both sets (of up to 6 objects) the same size.
❏ Matching picture sets 1-3 numerals.
❏ Making sets of 5-7 items.
❏ Identifying sets with more/less than 6.
❏ Making sets of 1-4 items.
❏ Matching given sets of up to 5 objects.
❏ Constructing sets from a given model.
❏ Giving one, and then one more, from a set of ten.

Sets Skills Checklist (Comparisons of Likenesses and Differences in Quantities)

- ❑ Counting by rote to 20.
- ❑ Identifying the larger of two numerals, 1-20.
- ❑ Recognizing and ordering cardinal numerals, 1-20, in sequence.
- ❑ Identifying numerals 1-20, when named in random order.
- ❑ Naming numerals, 11-20.
- ❑ Matching numerals 11-20.
- ❑ Matching sets of up to 10 objects with numerals.
- ❑ One-to-one-correspondence: counting 10 objects.
- ❑ Identifying ordinal positions, sixth through tenth.
- ❑ Solving simple word problems, using numerals, 1-10.
- ❑ Giving from 1-9 objects from a set of twenty, on request.
- ❑ Adding and subtracting numbers from 1-5.
- ❑ Solving simple word problems, using numerals, 1-5.
- ❑ Identifying ordinal positions, first through fifth.
- ❑ Counting by rote from 1-10.
- ❑ Identifying the larger of two numerals, 1-9.
- ❑ Identifying numerals 1-10, when named in random order.
- ❑ Recognizing and ordering cardinal numerals, 1-10, in sequence.
- ❑ Naming numerals, 1-3.
- ❑ Matching numerals 1-10.
- ❑ Giving from 1-4 objects from a set of ten, on request.
- ❑ One-to-one-correspondence: counting 4 objects.
- ❑ Solving simple word problems, using numerals, 1-3.
- ❑ Identifying ordinal positions, first, last and middle.
- ❑ Counting by rote to six.
- ❑ Giving 2 objects from a set of five or six, on request.
- ❑ One-to-one correspondence: counting 2 objects.
- ❑ Identifying ordinal position, first.
- ❑ Counting by rote to 3.
- ❑ Giving 1 object from a set of three, on request.
- ❑ Understanding the concept of "one."

Counting Skills Checklist

About Evaluation

Teachers are encouraged to periodically assess each child's progress through a skills checklist by observing behavior to determine whether or not learning has occurred. Teachers can then select activities and provide materials to help the child master the next rung on the checklist, as appropriate. This kind of evaluation is intended to help the teacher plan instruction, not to attach an age to a child's developmental level.

[1] Tobias, S. *Overcoming Math Anxiety.* New York: W.W. Norton & Company, Inc., 1978.

[2] Bringuier, J. *Conversations with Jean Piaget* (translated by Basia Miller Gulati). Chicago & London: The University of Chicago Press,1980.

[3] This book, and its companion publication *(More Blocks, Bears and Building Math Skills)*, are the predecessors to *Dice, Dominoes and Doing Math, A Collection of K-1 Activities Using Common Classroom Manipulables* .

[4] Gagne, Robert M. *Essentials of Learning for Instruction* : Hinsdale, Illinois: The Dryden Press, a division of Holt, Rinehart and Winston, Inc. 1975.

Activities for Three-Year-Olds

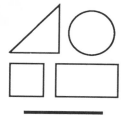

Filling Peg Shapes

What Children Learn
- To recognize simple plain geometric shapes (squares and triangles).
- To name simple plain geometric shapes.

Before You Begin
- Look at parquetry squares and triangles with the children. Allow the children to explore the shapes and make designs with them. Begin the following activities after the children have explored the shapes.

What Children Do
1. While the child watches, the adult makes a square or triangle on a pegboard with pegs in one color.
2. Identify the color together.
3. Identify the shape together.
4. The child fills in the shape on the pegboard with a different color peg. (See illustration.)
5. The child matches a parquetry shape with the pegboard.

More To Do
- Reproduce the shape from one pegboard onto another.
- Find other shapes to match the one on the pegboard.
- Make any kind of shapes from parquetry blocks.

Materials Needed
- Parquetry squares and triangles
- Large pegs in two colors
- Pegboards for large pegs

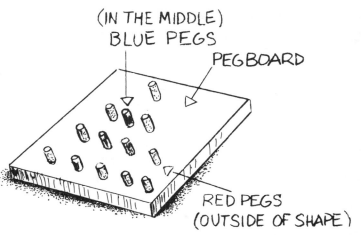

(IN THE MIDDLE)
BLUE PEGS

PEGBOARD

RED PEGS
(OUTSIDE OF SHAPE)

15

Making Simple Stick Shapes

What Children Learn
- To recognize squares and triangles.
- To name simple plain geometric shapes.
- To match items of the same color.

Materials Needed
- Craft sticks, some of which have been painted or dyed red, blue and green
- Glue and craft-stick-shapes lotto game

Before You Begin
- Make a craft-sticks-shapes lotto game by making lotto boards and draw cards with simple craft sticks shapes on them. Include the same shapes in a variety of colors to give the children practice matching colors as well as shapes.
- Identify the shapes with the children. Allow the children to play with the shape draw cards by sorting them and matching them in any way they wish.
- Practice matching colors with the children.

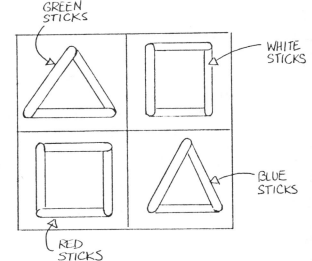

What Children Do
1. Use craft sticks and glue to make triangles by:
   - applying dots of glue to each end of three craft sticks;
   - joining them; and,
   - applying pressure until the glue sets.
2. Use craft sticks and glue to make squares by:
   - applying dots of glue to each end of four craft sticks;
   - joining them; and,
   - applying pressure until the glue sets.
3. Play craft-sticks-shapes lotto individually by drawing a card and finding a match on the board.
4. Lay craft sticks on the shapes on either the shapes lotto board or draw cards.

More To Do
- Play craft-sticks-shapes lotto in groups of two with both children completing their boards.
- Use the craft stick shapes to find squares and triangles in the classroom.

Shape Match Grab Bag

What Children Learn
- Shape discrimination skills.

Before You Begin
- Cut out large posterboard shapes of circles, squares, and triangles.
- Draw a circle, square or triangle on each of several paper bags.
- Place all posterboard shapes and blocks in a paper bag with no drawing ("the grab bag").

Materials Needed
- Circular, square and triangular blocks
- Posterboard and small paper bags
- Scissors and markers

What Children Do
1. Select a shape or block from the grab bag and match it to an object in the environment (e.g. circles to clocks).
2. Classify the blocks by shape, placing them in paper bags which have a drawing of the same shape.
3. Match blocks to posterboard shapes.

More To Do
- Use the blocks to build houses and towers.
- Give each child a posterboard shape. Ask children to form groups of three in which each shape is represented. Play "Shape Adventure:" a circle, square and triangle start out on a walk to find more shape friends. As each finds another with his own shape, that child joins the group.
- Give each child a posterboard shape. Have children travel around the whole classroom or playground until three groups are formed.

Circles Go Round and Round

What Children Learn
- To recognize circles.
- To make circles with simple tools.
- To understand that a circle is a continuous arc.

Before You Begin
- Give the children circles to play with before beginning the activities below. Encourage them to explore the perimeters of the circles.
- Make a hole in each end of each craft stick.
- Make a circle-maker for yourself and for each child, by fastening one end of a craft stick to the center of a piece of cardboard with a brass fastener;
- Model making a circle with the circle maker.
- Talk with the children about how a circle goes around and around, with no beginning and no end.

What Children Do
1. Make craft stick circles by:
   - poking a pencil point through the hole in the unfastened end of the craft stick; and,
   - rotating until a circle has been drawn on the cardboard.
2. Remove the brass fastener and craft stick, and identify the shape.

More To Do
- Use a lighter weight cardboard and have the children cut out the circles they make. They can use the cutout circles to locate other circles in the class.

Materials Needed
- A variety of circles in heavy materials
- Craft sticks
- Long brass fasteners
- Cardboard at least twice as tall and wide as the craft sticks are long.
- Pencils

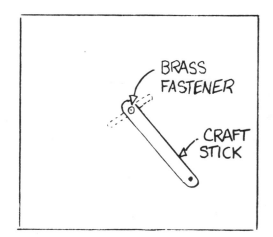

Color Sample Sort and Match

What Children Learn
- Color discrimination skills.
- Positional prepositions.

Before You Begin
- Cut the paper paint sample strips apart by color.
- Hold up the paint samples one at a time. Have children name each color. (Use primary and secondary colors only for this activity.)

What Children Do
1. Sort color samples into given categories: red and not red; blue and not blue; green and not green; yellow and not yellow.
2. Group paint samples by color (all the reds together; all the greens together, etc.).
3. Match color samples to the same color on the color wheel.
4. Match color samples to objects in the environment of the same color.
5. Match blocks to color samples by color.

Materials Needed
- Paper paint samples from a paint or hardware store
- Scissors
- A color wheel
- Colored blocks

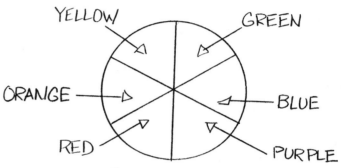

More To Do
- Find matching crayons by coloring on white paper and comparing the results with paint samples and the blocks.
- One child selects one of the color samples as a "favorite" color without letting others know what she has chosen. Others try to guess the child's color by asking questions about its position on the color wheel. Through questions (*Is it near red? Is it next to green? Is it opposite from blue?*) children will be able to identify the color while they practice using positional prepositions.

Clay Colors

What Children Learn
- How to classify objects by a common property (color or size).

Before You Begin
- Cut three paper circles (same colors as clay) and glue one on each box.
- Place boxes on the table.
- Shape three clay balls (large, medium and small) of each color.
- Place the clay balls randomly on a table.

What Children Do
1. Place all the small balls into the small box.
2. Repeat with medium and large balls and boxes.
3. Remove clay balls from the boxes and place them on the construction paper of the same color.
4. Remove clay balls from construction paper and place them in the box with a circle of the same color.
5. Remove clay from boxes and roll all the small balls into one big ball of each color.
6. Shape their own clay balls in various colors and sizes and repeat steps 1-3.

More To Do
- Allow time for experimentation and free play with the clay.

Materials Needed
- Clay (or play dough) in a variety of colors
- Colored construction paper, scissors and glue
- Three boxes (large, medium and small)

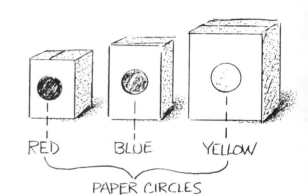

RED   BLUE   YELLOW

PAPER CIRCLES

20

Am I a Cube?

What Children Learn
- To sort three-dimensional objects by shape, size and weight.
- The concept of what it takes to roll.

Materials Needed
- Assorted sizes of cube-shaped blocks and boxes and spheres (balls)
- Red and blue posterboard sheets

Before You Begin
- Look at the shapes. Help the children to identify unique characteristics of cubes and spheres. Let the children generate the ideas with your guidance. Note the differences and name the shapes. Be sure to note that a cube has angles and faces and each face is identical. Note how spheres (balls) have no angles or faces but one continuous face.

- Decide which color of posterboard to use for cubes and which for spheres and place a sample of the shape on the posterboard.
- Discuss how to use hands to decide which of two objects is heavier. Model and give individual feedback so the children clearly understand the idea of heavier.

What Children Do
1. Sort the shapes onto the appropriate color of posterboard.
2. Sort the objects:
   - compare two objects by hand-weighing; and,
   - sort all the heavy solids from the others.
3. Sort the solids by size into big and little cubes and spheres (balls).
4. Decide which of the solids roll.

More To Do
- Build with cubes. *Can they build with spheres? Why not?*
- Build with cubes from models that are constructed by you or by other children.

Seesaw Math

What Children Learn
- To discriminate between heavy and light.

Before You Begin
- Talk about "heavy" and "light."
  Have children name things which are heavy
  (e.g. trucks, tables) and light (e.g. pencils,
  feathers).
- Weigh a variety of items on the balance scale.
  Decide which of each pair is heavy and
  which is light.
  Ask, *how does the balance scale show
  which is heavier?*
- Compare the weight of blocks to the weight
  of other small objects, using the balance scale.

Materials Needed
- Seesaw or teeter totter
- Balance scale
- Blocks and a variety of small objects
  (e.g. feathers, pencils)

What Children Do
1. Choose a playground partner.
2. Guess (predict) whose end of the seesaw will
   come closer to the ground.
3. Sit on opposite ends of the seesaw and check the prediction. (Pair the heaviest and
   lightest of each twosome with their counterparts from another twosome. Continue
   switching partners for more predictions and experimentation.)
4. Find a partner with whom they can balance the seesaw.
5. Weigh small classroom objects on the balance scale. Predict which is heavier and
   lighter and test predictions using the scale.

More To Do
- Take a nature walk. Pick up objects from the ground (stones, sticks, leaves, paper),
  and ask children to guess which is heavier. Have children take turns holding one of
  the objects in each hand and ask, *which "feels" heavier? Was your guess correct?*
- Discuss how a balance scale is like a seesaw (the heavier side goes down).

Longer or Shorter?

What Children Learn
- To compare objects to a specific unit.

Before You Begin
- Turn a large piece of posterboard horizontally and draw a line down the middle. One side is for objects longer than a craft stick and one side is for objects shorter than a craft stick. Glue a craft stick on each half of the chart. (See illustration.) Make the word "shorter," shorter than the craft stick and the word "longer," longer.
- Discuss how to decide if an object is longer or shorter than your foot. Apply the ideas to comparison with a craft stick.
- Help the children compare objects to a craft stick by laying a craft stick on or next to an object and deciding whether the object is longer or shorter than the stick.
- Decide on a symbol for the poster board chart to show longer-than-a-craft-stick and shorter-than-a-craft-stick.

Materials Needed
- Craft sticks
- Objects which are longer and shorter than a craft stick
- Posterboard, marker and glue

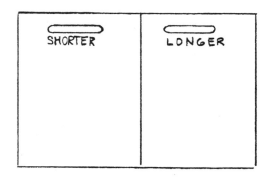

What Children Do
1. Compare an object to a craft stick using whatever strategy the children choose.
2. Decide if it is longer or shorter than the stick.
3. Put it on the posterboard on the longer side if it is longer or the shorter side if it is shorter.
4. Negotiate when there is a disagreement as to an object's size. (Disagreements may result if children are comparing different dimensions of an object.)

More To Do
- Measure pictures of objects with craft sticks.
- Make a summary chart by gluing a picture of each object on the appropriate side of another posterboard chart like that above. Some pictures may be available from catalogs or photographs.
- Provide objects which are equal in length to a craft stick and compare "longer," "shorter" and "equal to" a craft stick. Make a chart with three categories.

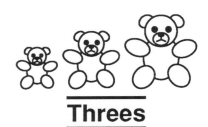

Rainbow Samples

What Children Learn
- Color discrimination skills.
- Concept of following a certain order.

Before You Begin
- Cut the paper paint sample strips apart by color.
  (Use primary and secondary colors for this activity.)
- Draw a rainbow with a black marker. Make a photocopy for each child.

What Children Do
1. Name the color of each paint sample.
2. Name the colors in the rainbow picture, starting with the top color.
3. Put paint color samples in the same order as they appear in the rainbow, from top to bottom.
4. Color the paper rainbow following the same order as that in the picture rainbow.

Materials Needed
- Paper paint samples from a paint or hardware store
- Scissors
- Picture of a rainbow
- Crayons or markers and paper

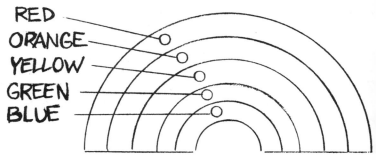

More To Do
- Play a singing game. Children take turns singing the following rhyme (to the tune of "The Farmer in the Dell"):

> My favorite color is _____.
> My favorite color is _____.
> Hi, ho the de-ri-o,
> My favorite color is _____.

Other children then point to something in the room which is the same as the color named.

Sandbox Relay

What Children Learn
- Measurement skills
- Gross Motor skills

Before You Begin
- Fill two buckets with sand.
- Set up two relay stations: place a sand-filled bucket at each starting line and an empty bucket, within comfortable running distance, at the finish line. Place a 1-cup measure at one starting line and a 1/2-cup measure at the other.

What Children Do
1. Line themselves up in size order, from shortest to tallest.
2. Every other child takes one step forward.
   (These children are team #1; others are team #2.)

Materials Needed
- Four sand buckets
- Four 1-cup and four 1/2-cup measuring cups

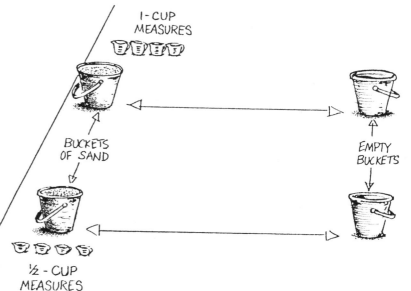

1-CUP MEASURES

BUCKETS OF SAND

EMPTY BUCKETS

1/2-CUP MEASURES

3. Run a relay race. First child in each line scoops sand from the filled bucket with the measuring cup, runs and pours it into the empty bucket, runs back to starting line and gives the cup to the next child. The first team to transfer all sand from the full to the empty bucket wins.
4. Have teams switch measuring cups for the next relay. Then ask, *why does the 1-cup line always finish first?*
5. Put the measuring cups in order from smallest to largest.
6. Repeat the relay, but this time place a 1-cup and a 1/2-cup measure at each starting line, allowing each runner to choose which measuring cup she will use when it's her turn to run.

More To Do
- Point to the 1-cup measure and the 1/2-cup measure and ask, *how many of these small cups (1/2-cup measure) of sand do you think it will take to fill the large cup (1-cup measure) ?*
- Place all buckets and measuring cups in the sand box for experimentation and free play.

How Many Bears in My House?   <u>Threes</u>

What Children Learn
- Counting items in two sets and in the combined set.
- Beginning addition skills.
- To recognize sets of two three, four, five and six without having to count.

Before You Begin
- Show the children the bear counters and the house.
- Make up a story with the children about blue bears visiting one room of the house, red bears visiting another room, etc.
- Count the set of bedroom bears, kitchen bears, living room bears, etc.
- Continue telling a story in which two rooms of bears join together. Count the combined set of bears. Allow the children to handle the dice and explore the surfaces. They can feel, sort and count them.

Materials Needed
- Bear counters in several colors
- Play house about the right size for the bear counters
- Large dice

What Children Do
1. Have each child:
   - place 1-5 bears into each room in the play house;
   - count the number of bears in each room of the house;
   - count the number of bears in a pair of rooms; and,
   - if advanced enough, count the bears in the whole house.
2. Roll two dice; then:
   - clap hands and count out loud, 1, 2, 3, etc., for each dot on the top of the die while placing a finger on the dot; and,
   - repeat the activity with new rolls of the die.
3. Have each child roll a handful of dice and match all dice which have the same number on top.

More To Do
- Have more advanced children recite the addition facts created such as, "A set of 2 bears and a set of 3 bears makes a set of 5 bears."

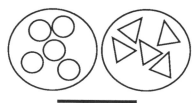

Checkers

What Children Learn
- Recognizing sets which are the same and different.
- Identifying sets of 1-4 items.
- How to make sets of 1-4 items.

Before You Begin
- Place 8 checkers of each color on the table.

What Children Do
Respond to questions about sets.
1. Teacher places an equal number of checkers in each color on the table and asks, *are these groups the same or different? How?* (Same number, different colors. Teacher places an unequal number of checkers in each color on the table and asks, *are these groups the same or different? How?* (Different number, different colors.)
2. Teacher makes two sets of three red checkers. Asks, *are these the same or different?* Child checks his response by counting and makes another set, just like the model. Teacher adds another checker to one set. Repeats question. Then asks, *how can we make the other two sets of checkers the same as this one?* Child adds one more red checker to the two smaller sets.
3. Repeat with sets of 1, 2 and 4 checkers.

Materials Needed
- Checkers

(RED CHECKERS)

"HOW CAN WE MAKE ALL THREE SETS THE SAME?"

More To Do
- Repeat activities using black checkers only. Then repeat again mixing colors to help children learn that number, not color, is the common factor in these sets.

Red Sets and Blue Sets

What Children Learn
- To compare sets, with up to three objects, and decide which has **more** or **less**.
- To estimate the number of objects in a given set and to use counters to determine whether or not the estimate is correct.

Materials Needed
- Blocks in red and blue

Before You Begin
- Discuss the idea of **sets**. We make sets by grouping things. We can create sets which have certain things in common. For example, we could put red blocks in one set and blue blocks in another. We could create a set which includes all blocks. That set includes the set of red and the set of blue blocks. We can compare sets by many attributes, such as color and size.

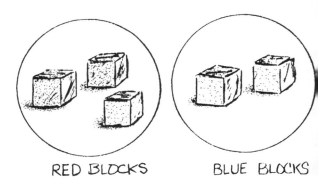

RED BLOCKS   BLUE BLOCKS

- Discuss the idea of **more**. Have the children select pairs of things and decide which has more, e.g. chairs or tables. *How do they decide?* Responses like, "it **looks** like **more** chairs than tables," are fine.
- Explain that they will now play a guessing game with the dice. Model the game below.

What Children Do
1. Give each child one to three dice in red and a different number of dice from one to three in blue.
2. Have each child:
    - guess whether he has more red or blue blocks;
    - stack the blocks in each color; and,
    - based upon height, decide which color has **more** blocks.
3. Ask, *was your guess correct?*

More To Do
- With more advanced children, distribute three groups of dice in three different colors. Have them guess which has **more** and **most**.

Simon Says..."Count to Three"

What Children Learn
- One-to-one correspondence.
- Following instructions and game rules.

Materials Needed
- Three blocks for each child

Before You Begin
- Explain that children will play "Simon Says..." using numbers.
- Have children stand up, facing "Simon" as in the traditional game.

What Children Do
Respond to Simon's instructions with their bodies. Examples:

> **Simon Says...**
> Close one eye.
> Close two eyes.
> Clap your hands three times.
> Raise two arms.
> Wiggle one leg.
> Point one foot.
> Hold up two fingers.

More To Do
- Give each child three blocks. Repeat the game, using blocks. Examples:

> **Simon Says ...**
> Balance one block on your head.
> Put two blocks on the floor under your chair.
> Give three blocks to your neighbor.

My Green Box

What Children Learn
- Pre-counting skills.
- Color names.
- Matching colors.
- Concept of more.

Before You Begin
- Select magazines with things in a color on which the children are working.
- Look at the counters with the children and name the colors.
- Match counters of the same color.
- Match counters to magazine pictures with the same color.

Materials Needed
- Green, red, yellow and blue counters
- Magazines with colored pictures
- A small box for each child
- White construction paper and glue

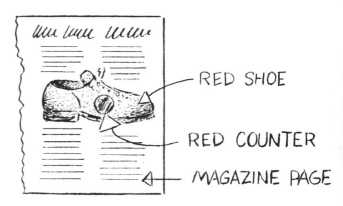

RED SHOE

RED COUNTER

MAGAZINE PAGE

What Children Do
1. Look through a magazine and find colors they like.
   Name colors they know.
2. Have each child:
   - select a color of counters;
   - pick out four or five counters in that color;
   - find four or five items of a matching color in magazines;
   - tear pictures from magazine; and,
   - make a box which has counters and magazine pictures in the same color.
3. Have each child count the green counters and count the green pictures. *How many of each are there?*
4. Have each child place a green counter on each green picture. *Are there **more** counters or pictures? Which one?*

More To Do
- Have the more advanced children match counters 1:1 with magazine pictures.
  How many counters did it take to equal the pictures?

Cookie Jar

What Children Learn
- One-to-one correspondence.
- Recognizing and ordering cardinal numerals 1-3 in sequence.

Before You Begin
- Attach a large numeral to each child's shirt with a safety pin.

What Children Do
1. Sit in a circle and respond to directions like:
   If you're a one, raise your hand.
   If you're a two, touch your toe.
   If you're a three, touch your knee.
2. Group themselves by numeral.
   All the one's, stand by the door.
   All the two's, sit on the floor.
   All the three's, sit in a chair.
3. Line themselves up in a repeating sequence of 1, 2, 3. (Or have children take turns lining others up in sequence.)

Materials Needed
- Numerals 1-3, each printed on a piece of newsprint, and safety pins
- Crackers and a "cookie jar"

More To Do
- At snack time, have children wash their hands carefully.
- Place crackers in a "cookie jar." Have children "sneak up" to the jar, one at a time, and "steal" the number of crackers described by his numeral. Have children switch numerals and repeat until all children have at least five crackers to eat for snack. (Children with more than five may share "extras" with the teacher.)

A Basic Board Game

What Children Learn
- To count the dots on a die.
- To use a die to play a game.
- A concept of top.

Before You Begin
- Prepare simple game boards with 12 or less stops on the path.
- Group children in pairs.
- Show children a die. Ask, *do you know what this is? Have you ever seen one before? Where? What is it used for?*
- Look at each side of the die.
- Roll the die and identify the top.
- Allow the children to play with the dice and compare the different sides of the dice.

What Children Do
1. Have one child in each pair:
   - roll a die;
   - identify the top of the die;
   - select the same number of counters as dots on the die top; and,
   - put the counters on the game board, one to a stop, in consecutive order.
2. Take turns checking each other.

More To Do
- As children develop more advanced skills, let them count the moves out using a marker.
- Use the game boards to invent their own games.

Materials Needed
- Game board with a path no longer than 12 stops
- From a block make a two-inch or larger die
  with two sides each having one, two and three dots
- Counters

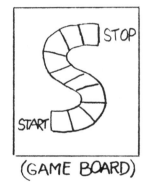

(GAME BOARD)

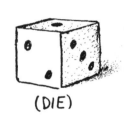

(DIE)

Checkers Stacks

What Children Learn
- That it is difficult to predict which of two equally possible options will occur
- Color discrimination; red and black

Materials Needed
- Checkers and a large sock

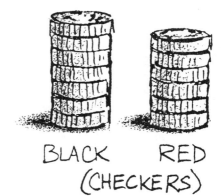

BLACK   RED
(CHECKERS)

Before You Begin
- Show children the checkers and ask them to name the colors of each. Have children point to other things in the room which are red or black.
- Stack an equal number of checkers in each color on the table and ask, *are these stacks the same or different? How?* (Same height, different colors.) Place all checkers into the sock and mix them up.

What Children Do
1. One child guesses what color checker she thinks she will get if she reaches into the sock, without looking.
2. Selects a checker from the sock and places it on the table.
3. States whether or not her prediction was correct.
4. Next child guesses, selects a checker and places it on the table (on top of the previous selection if the same color; next to previous selection, if different).
5. Game continues for eight rounds.
6. Have children look at the stacks of checkers which have been placed on the table. *Which stack is higher? Which color was picked from the sock most often?*
7. Return all checkers to the sock and play again.

More To Do
- Play the game using poker chips or counters of various colors (red, blue, green, etc.).

33

Picture Clock

What Children Learn
- That we can measure the passage of time by our activities.

Before You Begin
- Make a "picture clock" by drawing a large, paper circle on a piece of posterboard with magic marker. Cut it out.
- Divide the circle into pie-shaped wedges to represent each of your daily class activities.
- Make and attach movable hands to your "picture clock" with a paper fastener.

Materials Needed
- Large sheet of posterboard, a marker, scissors, paper fastener, magazines and paste

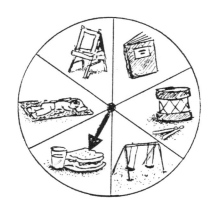

What Children Do
1. Cut pictures from magazines which represent each of their daily class activities (e.g. child sleeping for nap time, playground equipment for outdoor time, juice for snacktime).
2. Share their pictures and tell what activity it represents.
3. Select a class favorite in each category and put them in order from the first to the last thing they do at school each day.
4. Help you paste selected pictures (in clockwise order) on the large posterboard circle to make a picture clock.

More To Do
- Use the picture clock every day to note the passage of time and to make transitions from one activity to the next.

Activities for Four-Year-Olds

Making Simple Peg Shapes

What Children Learn
- To recognize simple plain geometric shapes (squares, triangles and rectangles).
- To name simple plain geometric shapes.
- To develop concepts of sides and angles.

Before You Begin
- Look at parquetry squares, triangles and rectangles with the children. Allow the children to explore the shapes and make designs with them. Begin the following activities after the children have explored the shapes.
- Look at the cardboard square with the children, and teach them to sing "I'm a Little Square" to the tune of "I'm a Little Teapot."

  I'm a little square,
  Filled with pride.
  Here is my angle.
  Here is my side.
  When I got all steamed up,
  Then I cried,
  "Tip me over to a new side."

Materials Needed
- Parquetry squares, triangles and rectangles
- Large pegs
- Pegboards for large pegs
- Square cut from cardboard

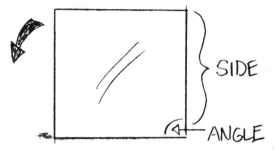

- Tip the square to rest on another side how the square looks the same no matter which side it rests upon.

What Children Do
1. While the child watches, the adult makes a square, triangle, or rectangle on a pegboard with pegs, then the child makes the same shape.
2. Identify the shape together.
3. Pairs of children can continue this activity on their own with some adult feedback.
4. Do another square activity:
   - make squares on the pegboard;
   - rotate them; and,
   - repeat the song learned above.

More To Do
- Provide a shape on a pegboard, and have the child make the same shape in a larger or smaller size.
- Match a shape to similar shapes in the classroom.

Making Stick Shapes

What Children Learn
- To recognize squares, triangles and rectangles.
- To name simple plain geometric shapes.
- Concepts of opposite sides and angles.

Before You Begin
- Cut some craft sticks in half and smooth any rough edges.
- Make models of each of the shapes for the children to follow.
- Identify the shapes with the children. Give the children craft sticks and allow them to make anything they wish with them. They should become familiar with the materials before beginning the activities which follow.
- Discuss the concept of shape sides and opposite sides with the children. Look at rectangles and note that opposite sides are the same length. Note also how the sides are placed like the sides of a square.

Materials Needed
- Craft sticks
- Glue

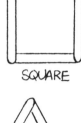

SQUARE

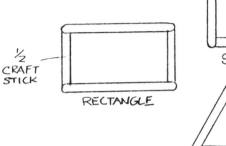

RECTANGLE

TRIANGLE

What Children Do
1. Use craft sticks and glue to make triangles from a model by:
   - applying dots of glue to each end of three craft sticks;
   - joining them; and,
   - applying pressure until the glue sets.
2. Use craft sticks and glue to make squares from a model by:
   - applying dots of glue to each end of four craft sticks;
   - joining them; and,
   - applying pressure and holding the sides to maintain the correct angles until the glue sets.
3. Use craft sticks and glue to make rectangles from a model by:
   - applying dots of glue to each end of two long, and two short, craft sticks;
   - joining them so the long sticks are opposite and lined up with each other, like the model, and the short sticks are opposite each other and lined up, like the model; and
   - applying pressure and holding the sides to maintain the correct angles until the glue sets.
4. Use the craft stick shapes to identify other triangles, squares, and rectangles in the classroom.

More To Do
- Use the craft stick shapes to identify other triangles, squares, and rectangles in the classroom.

Shape Match Treasure Hunt

What Children Learn
- Shape discrimination skills.

Before You Begin
- Cut out large posterboard shapes including circles, squares, rectangles, triangles and diamonds.

What Children Do
1. Match blocks to posterboard shapes.
2. Match posterboard shapes to objects and angles in the environment (e.g. circles to clocks, rectangles to window panes).
3. Go on a "Shape Match Treasure Hunt" outdoors and match posterboard shapes to objects and angles in the environment.

More To Do
- Play "I Spy." Children take turns saying, "I spy, with my little eye, something that looks like this shape..." (Child holds up a posterboard shape for all to see; others guess what the child "spies" which has the same shape.)

Materials Needed
- Blocks of varying shapes
- Posterboard, small paper bags
- Scissors, markers

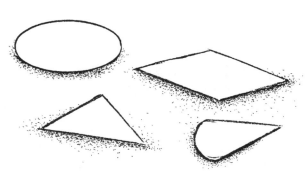

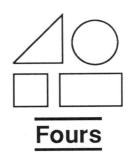

Circles

What Children Learn
- To recognize circles.
- To make circles with simple tools.
- Concept of a circle as a continuous arc.
- Concept that two circles with the same center have a special relationship. (They are concentric.)

Before You Begin
- Give the children circles to play with before beginning the activities below. Encourage them to explore the perimeters of the circles.
- Cut some whole craft sticks in half.
- Make a hole in each end of each craft stick. (See illustration.)
- Make a circle-maker for yourself by fastening one end of a craft stick to the center of a piece of cardboard with a brass fastener, and model making a circle with it.
- Make a circle with a compass and note how a circle goes around and around, with no beginning and no end.

Materials Needed
- A variety of circles in heavy materials
- Craft sticks, pencils and brass fasteners
- Cardboard at least twice as tall and wide as the craft sticks are long

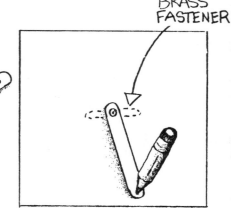

BRASS FASTENER

What Children Do
1. Make craft stick circle with a whole craft stick by:
   - fastening one end of a craft stick to the center of a piece of cardboard with a brass fastener;
   - poking a pencil point through the hole in the unfastened end of the craft stick; and,
   - rotating until she has a circle drawn on the cardboard.
2. Remove the brass fastener and craft stick (with adult help), and identify the shape with the children.
3. Have each child follow the directions above with the short craft stick to make a concentric circle.
4. Discuss how both circles have the same center. Note how their borders are the same distance from each other. (This is called concentric, because they both have the same center.)

More To Do
- Use a lighter weight cardboard and have the children cut out the circles they make. They can use the cut-out circles to locate other circles in the class.

Color Sample Lotto

What Children Learn
- Color discrimination skills.

Before You Begin
- Cut the paper paint sample strips apart by color.
- Make game boards by dividing the 8 1/2" x 11" piece of white paper into rectangles with a ruler and magic marker. (Be sure boxes are the same size as, or a bit larger than, the cut paint store samples.)
- Make multiple copies on card stock (at a local photocopy store).
- Paste color samples securely in each box on each card. (Use primary and secondary colors, black, white, brown and gray.)
- Laminate the lotto cards.

Materials Needed
- Enough paper paint samples from a paint or hardware store so that there are multiples of each color.
- White, unlined 8 1/2" x 11" paper
- A ruler, magic marker and paste

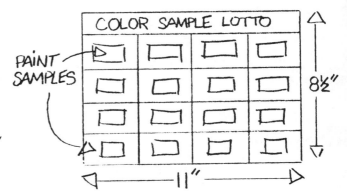

What Children Do
1. Say the names of the colors on the lotto board. Take turns drawing a sample from the sample pile and naming the color.
2. Match individual cut color samples to the colors on the lotto board, by placing the cut sample over the same color on the game board.
3. Select one of the colors and, by giving clues, help others guess the color. Sample clues: *Bobby is wearing it; it is near red on the lotto board; the color is in our rug.*

More To Do
- Match color samples to objects in the environment of the same color.

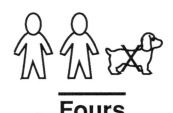

Clay Shapes

What Children Learn
- How to classify objects by a common property (color or shape).

Before You Begin
- Give each child some clay.

What Children Do
1. Fashion pieces of clay into a variety of shapes.
2. Group the clay pieces by color.
3. Invent other classification categories by responding to the question, *which clay pieces go together? Why?*
4. Group the clay pieces by categories identified in step #3.
5. Group the clay pieces by shape (e.g. round, flat, thin, thick, unusual, etc.).

Materials Needed
- Clay (or play dough) in a variety of colors

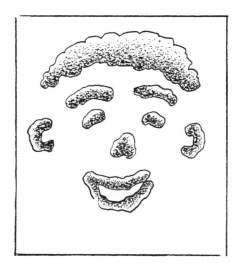

More To Do
- Fashion clay shapes to make "3-dimensional pictures" (e.g. faces, bodies, houses, flowers).
- Allow time for experimentation and free play with the clay.

42

Cubes, Spheres or Cones?

What Children Learn
- To sort three-dimensional objects by shape, size and weight.
- The concept of what it takes to roll.

Before You Begin
- Look at the shapes. Help the children to identify unique characteristics of each solid. Let the children generate the ideas with your guidance. Note the differences and name the shapes. Be sure to note that each face of a cube is identical whereas the rectangular solid has different faces. Note that the cone has points and edges, but spheres (balls) have none.
- Select a color of posterboard for each shape and place a sample of the shape on the sheet.
- Discuss how to use hands to decide which of two objects is heavier.
- Discuss how to use a balance scale to decide which of two objects is heavier.

Materials Needed
- Assorted sizes of blocks and boxes as in illustration 1
  (cubes, rectangular solids, and cones and spheres (balls))
- A different colored sheet of posterboard for each unique shape

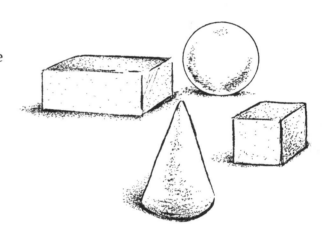

What Children Do
1. Sort the shapes onto the appropriate color of posterboard.
2. Sort the objects into those which are heavy and those which are light:
   - compare two objects by weighing in a hand;
   - sort using hands only; and,
   - for those objects which are close in weight, use a balance scale to decide.
3. Sort the solids by size into big and little.
4. Decide which of the solids roll.

More To Do
- Build with cubes, rectangular solids, and cones. *Can they build with spheres? Why not? Can they think of an idea for building with spheres?*
- Build with cubes from models which are constructed by you or by other children.
- Build with cubes from picture models. See illustration 2.

Seesaw, Marjorie Daw

What Children Learn
- Making and checking simple predictions.
- Weight measurement skills.

Before You Begin
- Take children outdoors and encourage free play on the seesaw or teeter totter.

What Children Do
1. Have children chant this rhyme as they go up and down on the seesaw:

> Seesaw, Marjorie Daw,
> One can weigh more than the other.
> Seesaw Marjorie Daw,
> My sister weighs more than my brother.
>
> (Alternate last line:
> "My father weighs more than my mother"
>
> Repeat substituting "less" for "more" in the rhyme.

Materials Needed
- Seesaw or teeter totter
- A balance scale and blocks of various weights

More To Do
- Place blocks on a table.
- Show children the balance scale and discuss how it works like a seesaw.
- Sort the blocks into two categories; heavy and light.
- Try to balance the scale using blocks (e.g. two blocks of the same size will balance the scale).
- Experiment in response to teacher questions (e.g. *what happens if we add another block to one side? Do two small blocks weigh the same as one large block?*).
- Experiment on their own by trying to balance various combinations of blocks.

How Many Craft Sticks?

What Children Learn

- To measure objects in a specific unit.
- To develop a concept of rounding off.

Before You Begin

- Discuss strategies for measuring things. Measure furniture in the classroom by using your feet. *What do you do if there is a partial foot? How do you decide whether or not to add a foot? Why do different people get different answers for the same object?*
- Help the children mark the middle of each craft stick.
- Show the children how to measure objects using craft sticks. Lay an appropriate number of craft sticks on the object. Count the number of craft sticks. To estimate a partial craft stick use the following rules: If the end of the object comes to the middle or further on the stick, count the stick as another whole stick. If the end of the object doesn't come to the middle, don't count the craft stick.

Materials Needed

- Craft sticks
- Numbers from one through five written on cards
- Markers
- Objects which are no more than 5 craft sticks in any dimension

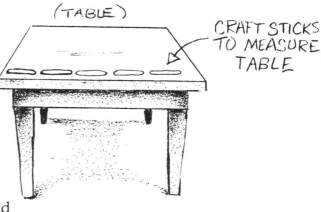

(TABLE)

CRAFT STICKS TO MEASURE TABLE

What Children Do

1. Use craft sticks to measure various objects in the classroom and tape a numeral on the object that corresponds to the number of craft sticks which the object measured.
2. Negotiate when there is a disagreement as to a measurement. (Disagreements may result if children are measuring different dimensions of an object.)
3. Discuss why the results of measuring with a craft stick are more consistent than when using our feet. (Feet are different sizes.)

More To Do

- Measure longer objects.
- Make a chart with a picture of each object and the measurement next to it. Some pictures may be available from catalogs.
- Use something other than a craft stick or foot to measure the same objects. What happens? How do the craft stick measurements compare to the new measuring device?

Color Wheel Samples

What Children Learn
- Color discrimination skills.
- Concept of following a certain order.

Before You Begin
- Cut the paper paint sample strips apart by color. (Use primary and secondary colors for this activity.)
- Draw a color wheel with a black marker. Make a photocopy for each child.

What Children Do
1. Name the color of each paint sample.
2. Name the colors in the color wheel.
3. Put paint color samples in the same order as they appear in the color wheel.
4. Color the paper color wheel with crayons or markers, following the same order as that in the color wheel.
5. Paint a color wheel, following the color sequence, using tempera paints

Materials Needed
- Paper paint samples from a paint or hardware store
- Scissors
- A color wheel
- Crayons or markers and paper
- Tempera paints, brushes and paper

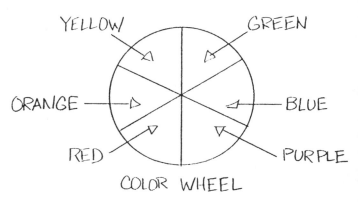

More To Do
- Experiment with color by mixing tempera paints, (primary and secondary colors only; no black or white) to make new colors.

Sandbox Math

What Children Learn
- How to discriminate between objects of different weights.
- Making and checking simple predictions.

Before You Begin
- Fill the 6 opaque containers with differing amounts of sand.

What Children Do
1. Have each child in turn pick up any two cans of sand. Ask, *which is heavier?* Have child check his prediction by placing the cans on the balance scale.
2. Have each child arrange the cans of sand from lightest to heaviest by picking them up and making a judgement based on how they feel. Each checks prediction using the balance scale.
3. Show children the 1-cup measure and the 1/2-cup measure. Ask questions like:

   - How many of these small cups of sand do you think it will take to fill the large cup?
   - What do you think would happen if you poured that large cup of sand into one dish of the scale and a small cup of sand into the other dish?

Materials Needed
- Sand, 6 clean, empty, opaque containers of equal size (e.g. canned foods), a balance scale

CANS OF SAND

LIGHTEST — HEAVIEST

More To Do
- Place measuring cups and the balance scale outdoors in the sandbox (or indoors at the sand table) for experimentation and free play.

How Many Bears in Each Set?

What Children Learn
- Identifying collections of items as sets and combining sets.
- Beginning addition skills.
- Matching sets of twos, threes, etc.

Before You Begin
- Take a handful of red bears and count them. CAUTION: Select a quantity of bears appropriate to your children's counting skills. Talk about the set of "red" bears. Ask, *how many are in the set of red bears?*
- Take a handful of blue bears and count them. Talk about the set of "blue" bears. Ask, *how many are in the set of blue bears?*
- Count all the bears. Talk about the set of "all" bears. Ask, *how many are in the set of all bears?*
- Repeat several times.
- Allow the children to handle the die and explore its surfaces.

Materials Needed
- Red and blue bear counters (or any two colors)
- Two boxes, each marked with a color to match one of the bear colors
- A very large die

What Children Do
1. Have each child:
   - sort the bears by color into the appropriate boxes;
   - take a handful of red bears from the appropriate box and count them;
   - take a handful of blue bears and count them;
   - count all the bears; and,
   - repeat the activity with new handfuls of bears.
2. Count bears by:
   - selecting a few red bears from the appropriate box and a few blue bears and placing them all far apart on the floor in a long line;
   - marching next to the bears, clapping and counting out loud, 1, 2, 3, etc., for each bear when next to it; and,
   - repeating the activity with new sets of dice.
3. Have each child roll a very large die (in one color) and make a set of bears with the same number on the top of the die.

More To Do
- For more advanced children provide more bears.

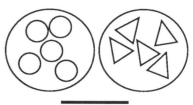

Checker Sets

What Children Learn
• Identifying sets with more or less than six.
• How to make sets, and subsets, of 1-7 items.

Materials Needed
• Several sets of checkers
• Checker boards

Before You Begin
• Arrange several sets of 1-7 checkers on the table.
  At first, make the sets with one color only,
  so children will concentrate on number, not color.

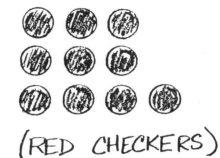

(RED CHECKERS)

What Children Do
1. Identify sets which contain less than six checkers,
   from sets you have made on a table.
2. Repeat, using sets of mixed colors. Help children
   understand that number, not
   color, is the identifying property of the set.
3. Divide sets into subsets. For example: place a set of 5 checkers (3 red, 2 black) on the
   table and ask, *how many checkers are in this group?* Divide checkers into subsets
   (by color) and ask questions like, *are there still the same number of checkers?* (Yes, 3
   red and 2 black still total 5 checkers.)

More To Do
• Create their own sets, and subsets, using 1-7 checkers in each set.
• Play checkers on the checker board.

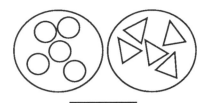

Guess How Many in a Set

What Children Learn
- To compare sets, with up to four objects, and determine which has **more** or **less**.
- To estimate the number of objects in a given set.

Before You Begin
- Discuss the idea of **sets.** We make sets by grouping things. We can create sets which have certain things in common. For example, we could put blocks in one set and bear counters in another. We could create a set which includes all red counters. That set includes part of the set of blocks and part of the set of bears. We can compare sets by many attributes, such as color and size.

Materials Needed
- A variety of counters (bears, round, blocks) in two colors
- 2" dice (Make them yourself out of blocks.)

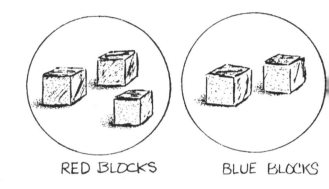

RED BLOCKS   BLUE BLOCKS

- Show children several sets of counters with different numbers in them. Ask, *are these sets the same or different? How?*
- Talk about **more**. Compare a variety of groups in the class, e.g. boys and girls. Guess which has **more**. Develop strategies for comparing.
- Explain that they will now play a guessing game with the sets of counters. Talk about "guessing." Model the game below.

What Children Do
1. With two unequal sets of different counters:
   - guess which set has **more**; and,
   - match the set members one-to-one, and decide which set has **more** members;
2. With two sets of block counters:
   - guess which set of dice has **more**; and,
   - stack the blocks from each set, and, based upon height, decide which set has **more** blocks;
3. Have each child:
   - roll two of the 2" dice until they have different numbers of dots on top;
   - guess which die has **more** dots on top;
   - match counters to each of the dice;
   - stack them to determine which die has more dots on top; and,
   - repeat process with other rolls of the dice.
4. Ask, *was your guess correct*?

More To Do
- Repeat all of the above, looking for the set or die which has **less**.

50

Simon Says..."Count to Ten"

What Children Learn
- One-to-one correspondence.
- Following instructions and game rules.

Materials Needed
- Ten blocks for each child

Before You Begin
- Explain that children will play "Simon Says..." using numbers.
- Have children stand up, facing "Simon" as in the traditional game.

What Children Do
Respond to Simon's instructions with their bodies. Examples:

Simon Says...
> Close one eye.
> Close two eyes.
> Clap your hands three times.
> Tap your foot four times.
> Turn around two times.
> Touch your toes three times.
> Hold up six fingers.
> Touch your nose two times.
> Hop on one foot four times.
> Jump five times.
> Hug your neighbor one time.
> Tap your head three times.

More To Do
- Give each child ten blocks. Repeat the game, using blocks. Examples:

Simon Says ...
> Balance one block on your head.
> Put two blocks on the floor under your chair.
> Give three blocks to your neighbor.
> Build a tower with six blocks.
> Make a train with seven blocks.
> Build a house with nine blocks.

My Green Picture

What Children Learn
- How to count.
- Color names.
- To use counting for non-related activities.

Before You Begin
- Select magazines with things in a color on which the children are working.

What Children Do
1. Look through a magazine and find colors they can name. Select colors they like.
2. Sort out the green counters from the non-green counters, then:
   - cut or tear green things from magazine pictures and glue them onto a piece of construction paper;
   - match a green counter to each green item cut from the magazine;
   - count the number of green things they have glued onto the construction paper;
   - count the number of green counters *(are there the same number of green counters as magazine pictures?)*; and,
   - do the same for the red, yellow and blue counters at another time.

Materials Needed
- Green, red, yellow and blue counters
- Magazines with colored pictures
- Scissors
- White construction paper and glue

More To Do
- Provide white construction paper which has a chart with horizontal lines laid out on it, and have the children glue green things on one line of the chart, red on another, yellow on another and blue on another. (See illustration.) Note which color has more items.

Who Stole the Cookie?

What Children Learn
- One-to-one correspondence.
- Recognizing and ordering cardinal numerals 1-10 in sequence.

Before You Begin
(Note: This is a good game to play during outdoor time in the playground.)
- Attach a large numeral to each child's shirt with a safety pin. (Ten children at a time can participate. Give each child a different numeral every time she plays the game.)
- Line children up in sequential order in a circle. (Once children have had practice recognizing the numerals, have them "line themselves up" in sequence, without adult help, for future games.)

Materials Needed
- Numerals 1-10, each printed on a piece of newsprint, and safety pins
- Crackers and a "cookie jar"

What Children Do
Clap hands in unison to create a rhythm as they chant:

All:	Who stole the cookie from the cookie jar?
Teacher:	Number one stole the cookie from the cookie jar.
Child #1:	Who, me?
All:	Yes, you!
Child #1:	Couldn't be!
All:	Then, who?
Child #1:	Number two stole the cookie from the cookie jar.
Child #2:	Who, me?
All:	Yes, you!

(Game continues around the circle in sequential, not random, order. Child #10 replies that #1 stole the cookie and the cycle begins again.)

More To Do
- At snack time, place crackers in a "cookie jar." Have children "sneak up" to the jar, "steal" a cracker and take it to their seats for snack. Before eating, have children count themselves and then count the crackers. *Are here the same number of children as crackers?* Refill the "cookie jar" and repeat until all children have "stolen" and counted crackers. Eat and enjoy!

A Board Game

What Children Learn
- To count the dots on a die.
- To use dice to play a game.

Materials Needed
- Game board with a path and markers
- Large dice
- Counters

Before You Begin
- Prepare simple game boards with 15-20 steps on the path.
- Group children in pairs.
- Show children a die. Ask, *do you know what this is?*
  Have you ever seen one before? Where?
  What is it used for?
- Roll a die and identify the top. Repeat several times.
- Model rolling the dice, placing a counter on the path for each dot on the top of the die, and moving a marker to the position of the last counter.
- Allow the children to play with the dice and compare the different sides of the dice.

What Children Do
1. Have one child in each pair:
   - roll a die to see how far to move on the board;
   - place counters on the path for each dot showing on the top of the die; and,
   - move the marker to the space on the path indicated by the last counter.
2. Take turns until they all reach the end of the path.

More To Do
- Count the number of dots on the top of the die and move the marker the correct number of spaces on the board.

Heads or Tails?

What Children Learn
- How to document results on a simple bar graph

Before You Begin
- Draw a model for a simple bar graph. (See illustration.)
- Give each child a bar graph and crayons or markers.
- Place all checkers in the sock and mix them up.

Materials Needed
- Several sets of checkers and a large sock
- Paper and markers or crayons

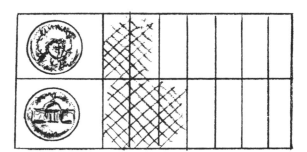

What Children Do
1. Select a checker from the sock
2. Guess whether the checker will come up "heads" or "tails" when flipped.
3. Flip a checker and note whether or not prediction was correct.
4. Color the box in the bar graph for "heads" or "tails," whichever was flipped.
5. Continue play for ten rounds.
6. Look at the bar graph. *Which came up more often, "heads" or "tails?"*
7. Play ten more rounds. *Are the results the same or different?*

More To Do
- Make pie charts using the information gathered above.
- Compare the bar graphs to the pie charts.

Shadow Clock Sundial

What Children Learn
- That the sun helps us measure the passage of time.

Materials Needed
- Posterboard, scissors and tape
- Battery-operated clock

Before You Begin
- Cut a large circle out of posterboard.
- Cut a triangular piece of posterboard (one side of which is half the diameter of the circle). Fold and tape the triangle to the circle so it stands up straight and casts a shadow. (See illustration.)

What Children Do
Go outside at three different times on a sunny day (early morning, noon and late afternoon, if possible). Each time you go out ...

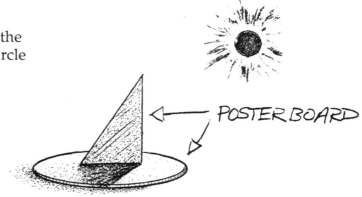

POSTERBOARD

1. Have each child stand in the same spot. Examine their shadows.
2. Look at the Shadow Sundial and mark the end of the shadow with a stick. *Has the shadow moved? Why?*
3. Compare the Shadow Sundial to the clock. *How are they similar? How are they different?*

More To Do
- Note the sun's position in the sky each time you go out. *What effect does this have on the shadows? Are shadows longer at some times of day that at others? Why?*

Activities for
Five-Year-Olds

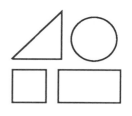

Making Advanced Peg Shapes

What Children Learn
- To recognize simple plain geometric shapes (squares, triangles, rectangles and diamonds).
- To name simple plain geometric shapes.
- To play a game.

Before You Begin
- Look at parquetry squares, triangles, diamonds and rectangles with the children. Allow the children to explore the shapes and make designs with them. Begin the following activities after the children have explored the shapes.

Materials Needed
- Parquetry squares, triangles, diamonds and rectangles
- Pegs
- Pegboard
- Rectangles and a square cut from construction paper

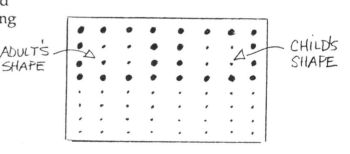

What Children Do
1. While the child watches, the adult makes a square, triangle, diamond or rectangle on a pegboard with pegs and the child makes the same shape.
2. Identify the shape together.
3. Pairs of children can continue this activity on their own with some adult feedback.
4. Play a shapes version of musical chairs:
   - use enough chairs for each child;
   - put a square on one chair and rectangles on the rest (see illustration);
   - while music plays chant, "rectangles and squares, musical chairs;"
   - when the music stops each child rushes to a chair with a rectangle on it;
   - the child left with the chair with the square on it, is out;
   - remove one chair with a rectangle;
   - rearrange the others and continue; and,
   - play until only one child remains.
5. Repeat playing shapes musical chairs using triangles and a diamond and squares and a diamond.

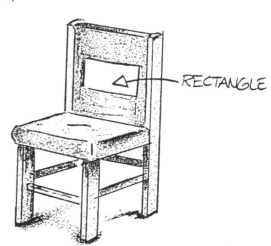

More To Do
- Provide pictures of the shapes on the pegboards labeled with the shape name. Have the children reproduce the shapes from the pictures.

Stick Shapes and Angles

What Children Learn
- To recognize simple squares, triangles, rectangles and diamonds.
- To name simple plain geometric shapes.
- Concepts of opposite sides, angles, right angles and parallel.

Materials Needed
- Craft sticks and glue
- Saw or serrated table knives and sandpaper

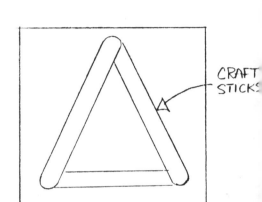

CRAFT STICKS

Before You Begin
- Identify the shapes with the children. Give the children craft sticks and allow them to make anything they wish with them. They should become familiar with the materials before beginning the activities which follow:
- Discuss the concept of opposite sides with the children. Look at rectangles and note that opposite sides are the same length. Note also how the sides are placed like the sides of a square at right angles to each other. This makes the opposite sides parallel to each other.
- Note that a diamond has four equal sides and opposite sides are parallel to each other, but the angles are not right angles. Ask, *"what shape would it be if you made the angles look like this?* (Make a right angle.)

What Children Do
1. Use craft sticks and glue to make triangles and squares from a model by:
   - applying dots of glue to each end of three or four craft sticks;
   - joining them; and,
   - applying pressure until the glue sets. (Maintain correct angles for the square.)
2. Use craft sticks and glue to make rectangles from a model by:
   - using saws or serrated table knives to cut one craft stick in half (**SUPERVISE CAREFULLY**);
   - smoothing any rough edges with sandpaper,
   - applying dots of glue to each end of two long, and two short, craft sticks;
   - joining them so the long sticks are opposite and lined up the same distance apart (parallel) from each other and the short sticks are parallel to each other; and,
   - applying pressure and holding the sides to maintain the correct angles until the glue sets.
3. Use craft sticks and glue to make diamonds from a model by:
   - applying dots of glue to each end of four craft sticks;
   - joining them, but not at right angles; and,
   - applying pressure and holding the sides to maintain the correct angles until the glue sets.

More To Do
- Use the craft stick shapes to identify other triangles, squares and rectangles.

Shape Match Bingo

What Children Learn
• Shape discrimination skills.

Before You Begin
• Make game boards by dividing
  an 8 1/2" x 11"
  piece of white, unlined paper into
  12 squares with a ruler and magic marker.
• Make multiple copies on card stock (this can be done
  at a local photocopy store).
• Draw shapes in each box on each card. (Be sure no
  two cards are exactly the same.)
• Laminate the bingo cards.

What Children Do
1. Play "Shape Bingo."
2. Select shape blocks randomly from a paper bag
   (replacing each as it is used).
   Have children place a counter or poker chip on a
   corresponding shape on his board.
3. Continue play until each child has covered his board.

Materials Needed
• Blocks of varying shapes
• Scissors and markers
• Unlined white paper
• Counters or poker chips

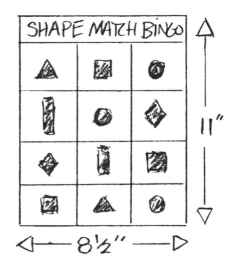

More To Do
• Make multiple copies of the bingo board, draw a shape in each box, cut out the
  boxes and laminate. Use these "shape cards" and a bingo board to play "Shape Lotto."

Circle Clocks

What Children Learn
- To recognize circles.
- To make circles with simple tools.
- Concept of a circle as a continuous arc.
- Concept of a clock.

Before You Begin
- Give the children circles with craft sticks fastened in the middle to play with before beginning the activities below. Encourage them to explore the movement of the craft sticks in relationship to the circles.
- Cut some whole craft sticks in half.
- Make a hole in each end of each whole craft stick and at the square end of the cut stick.
- Make a circle-maker for yourself by fastening one end of a craft stick to the center of a piece of cardboard with a brass fastener, and model making a circle with it.
- Talk with the children about how a circle goes around and around, with no beginning and no end.

Materials Needed
- A variety of circles in heavy materials
- Craft sticks and brass fasteners
- Cardboard at least twice as tall and wide as the craft sticks are long
- Clocks

What Children Do
1. Make craft stick circles by:
   - fastening one end of a craft stick to the center of a piece of cardboard with a brass fastener;
   - poking a pencil point through the hole in the unfastened end of the craft stick; and,
   - rotating until she has a circle drawn on the cardboard.
2. Remove the brass fastener and craft stick, and identify the shape with the children.
3. Make clocks out of their circles by:
   - making clock face numerals on the cardboard with magic markers (for children who cannot make their own, draw dotted numerals for them to trace around);
   - using a brass fastener to put a whole stick and a half stick on the cardboard through the center hole; and,
   - rotating the hands to times they can identify.

More To Do
- Reproduce times with their clocks from paper or real clock models.
- More advanced children can reproduce times with their clocks from instructions given verbally or in writing.

Color Sample Bingo

What Children Learn
- Color discrimination skills.

Before You Begin
- Cut the paper paint sample strips apart by color.
- Make game boards by dividing the 8 1/2" x 11" piece of white paper into squares with a ruler and magic marker. (Be sure boxes are the same size as, or a bit larger than, the cut paint store samples.)
- Make multiple copies on card stock (at a photocopy store).
- Paste color samples securely in each box on each card. (Use varying hues from each color family.) Make each card slightly different from the others through color choices and arrangement.
- Laminate the bingo cards.
- Place one sample of each color into a paper bag.

Materials Needed
- Enough paper paint samples from a paint or hardware store so that there are multiples of each color
- White, unlined 8 1/2 x 11" paper
- A ruler, magic marker and paste
- Paper bag
- Teddy Bear counters or chips

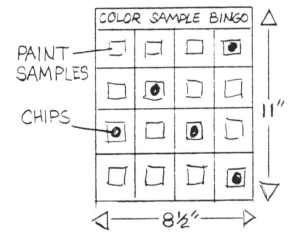

What Children Do
1. Group paint samples by color family (e.g. all the blues together).
2. Play bingo.
   - Place one of each color sample into a paper bag.
   - Have one child select a color sample from the bag.
   - Others place a counter or poker chip over the corresponding color on their boards.
   - Put selected sample back in the bag and draw again.

More To Do
- Match color samples to objects in the environment (including walls) of the same hue.

Clay Weights

What Children Learn
- How to classify objects by a common property (weight).

Before You Begin
- Place clay and balls on a table.

What Children Do
1. Make clay balls of varying sizes.
2. Pick up a clay ball in each hand and guess which is heavier.
3. Check guess using balance scale.
4. Arrange the clay balls from lightest to heaviest.
5. Sort all balls (including clay balls) into two categories; heavy and light.
6. Note that size is not necessarily related to weight (e.g. the foam ball may be bigger than the gum ball, but may not be heavier on the balance scale.)

Materials Needed
- Clay (or play dough) and a balance scale
- A variety of balls (e.g. ping pong ball, softball, foam ball, basketball, gum ball)

More To Do
- Shape two equal-sized clay balls. Check equality of weight using the balance scale. Add or remove clay until the balls balance. Flatten one of the balls. Ask children, *which clay shape has more?*
- Re-shape the flattened ball by rolling into a stick shape. Ask children, *which clay shape has more?*

(NOTE: Accept all responses to questions without trying to "teach" the "correct" answer. In his experiments Piaget found that very young children perceive "more" clay in the stick shape because it is longer. Later they perceive "less" clay because it is "thinner." It is not until elementary school that children truly understand that altering the shape of an object does not not change its weight or volume.*

* Bringuier, J. Conversations with Jean Piaget (translated by Basia Miller Gulati). Chicago & London: The University of Chicago Press, 1980 (pps. 31-33).

What's My Shape?

What Children Learn
- To sort three dimensional objects by shape, size and weight.
- The concept of what it takes to roll.

Before You Begin
- Look at the shapes. Help the children to identify unique characteristics of each solid. Let the children generate the ideas with your guidance. Note the differences and name the shapes. Be sure to note that each face of a cube is identical whereas the rectangular solid has different faces. Note that the cone is smooth and the pyramid has flat surfaces that meet. Note how spheres (balls) have no angles at all.
- Select a color of posterboard for each shape and place a sample of the shape on the sheet.
- Discuss how to use hands to decide which of two objects is heavier.
- Discuss how to use a balance scale to decide which of two objects is heavier.

Materials Needed
- Assorted sizes of blocks and boxes: cubes, rectangular solids, pyramids, cones and spheres (balls)
- A different colored sheet of posterboard for each unique shape
- Balance scale

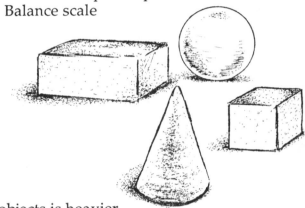

What Children Do
1. Sort the shapes onto the appropriate color of posterboard.
2. Sort the objects into those which are heavy and those which are light:
   - compare two objects by hand-weighing; and,
   - for those objects which are close in weight, use a balance scale to decide.
4. Decide which of the solids roll.
5. Build structures with the shapes, both freeform and from models.

More To Do
- For more advanced children provide unassembled cubes and let them assemble them with glue or tape. They can paint the cubes after assembly with tempera paints, or color them with crayons before assembly.
- Have the more advanced children use weights on the balance scale to weigh the objects and decide which weigh more than others.

Balance Scale Math

What Children Learn
- Making and checking simple predictions.
- Weight measurement skills.

Before You Begin
- Place a variety of small objects on a table.
- Show children the balance scale and discuss how it works like a seesaw.

What Children Do
1. Choose two objects at random and guess which is heavier.
2. Compare the actual weights, using the balance scale.
3. Sort the objects into two categories; heavy and light.
4. Line up the objects in order from lightest to heaviest.

More To Do
- Go outside and play on the playground.
- Sort themselves into two categories; heavy and light.
- Predict which of two seesaw partners is heavier.
- Find a partner with whom they can balance the seesaw.
- **SUPERVISE CAREFULLY.** Have light and heavy children experiment with sliding forward to see how this action affects the balance.

Materials Needed
- Clay (or play dough) in a variety of colors

Craft-Stick Rulers

What Children Learn
- To measure objects in a specific unit.
- How a ruler is made.
- To use a ruler.
- To develop a concept of rounding off.

Materials Needed
- Craft sticks and glue
- Markers
- Long narrow pieces of cardboard or wooden slats about 2" by 36"

Before You Begin
- Help the children mark the middle of each craft stick.
- Draw a straight line 3/4" from the edge of the cardboard wooden slat.

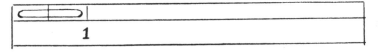

- Discuss strategies for measuring things. Measure furniture in the classroom by using your feet. What do you do if there is a partial foot? How do you decide whether or not to add a foot? Why do different people get different answers for the same object?
- Show the children how to measure objects using craft sticks. Lay an appropriate number of craft sticks on the object. Count the number of craft sticks. To estimate a partial craft stick use the following rules: If the end of the object comes to the middle or further on the stick, count the stick as another whole stick. If the end of the object doesn't come to the middle, don't count the craft stick.

What Children Do
1. Use craft sticks to measure various objects in the classroom and tape a numeral on the object that corresponds to the number of craft sticks which the object measured.
2. Negotiate when there is a disagreement as to a measurement. (Disagreements may result if children are measuring different dimensions of an object.)
3. Make craft-stick-rulers. Glue craft sticks end-to-end on the line on the cardboard or wooden slat. From left to right, number the end of each craft stick: 1, 2, 3, 4, etc. (See illustration 1.)
4. Use the rulers to measure objects previously measured. Is the measurement the same as with individual craft sticks?

More To Do
- Measure longer objects.
- Make a chart with a picture of each object and the measurement next to it. Some pictures may be available from catalogs, or photograph the objects the children will be measuring ahead of time. (See illustration 2.)

OBJECT	LEGNTH
	16

Illustration 2

Color Sample Series

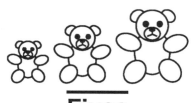

What Children Learn
• Color discrimination skills.

Before You Begin
• Cut the paper paint sample strips apart by color.
• Give each child five or more samples in the same color family (e.g. all the blues).

What Children Do
1. Place paint samples from one color family (e.g. all the blues) in order from lightest to darkest.
2. Switch samples with someone who has a different color family; mix samples up and repeat.

More To Do
• Experiment with color by mixing tempera paints, including white, to make varying hues in each color family.
• Paint a picture, using only one color plus white.

Materials Needed
• Paper paint samples (in varying gradations of color) from a paint or hardware store
• Tempera paints, brushes and paper

PAINT STORE SAMPLES

Sandbox Fractions

What Children Learn
- Making and checking simple predictions.
- Measurement and fractions skills.

Before You Begin
- Draw and cut out cup measures of three sizes (1-, 1/2-, and 1/4-cup) from construction paper. (Use a different color for each size cup.)

What Children Do
1. Place construction paper cups in order from smallest to largest.
2. Place the plastic measuring cups in order from smallest to largest.
3. Make predictions based on your questions. For example: point to the 1-cup measure and one of the fractional measuring cups (1/2, 1/4) and ask:
    - *How many of these small cups of sand do you think it will take to fill the large cup?* (Have child select construction paper cutouts for his estimate; then he can check his prediction by pouring sand from the fractional measuring cup into the 1- cup measure.
    - *What do you think would happen if you poured that large cup of sand into one dish of the balance scale and a small cup of sand into the other dish?*
4. Experiment by trying to balance the scale using various combinations of sand-filled measuring cups. Pose questions like: *how many 1/4-cup measures will balance one 1/2-cup measure? The 1-cup measure?* Repeat with other combinations.

Materials Needed
- Colored construction paper and scissors
- Sand, a balance scale and a scale with numbers indicating pounds
- Plastic measuring cups (1-cup, 1/2-cup and 1/4-cup) and a sand bucket

"HOW MANY SMALL CUPS WILL FILL THE LARGE CUP?"

More To Do
- Use the measuring cups to pour sand into the sand bucket and weigh the bucket on the scale with numbers. *What happens?* (The more sand in the bucket, the higher the numbers on the scale.)
- Place the measuring cups and the balance scale outdoors in the sandbox (or indoors at the sand table) for experimentation and free play.

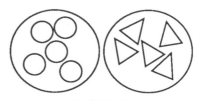

Bears, Bears, How Many Bears?   <u>Fives</u>

What Children Learn
- Counting items in two sets and in the combined set.
- Beginning addition.
- To recognize sets of two, three, four, five and six without having to count.

Materials Needed
- Bear counters and dice
- Individualized worksheets as below and markers

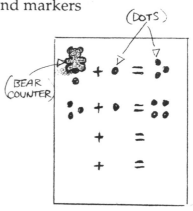

Before You Begin
- Make individualized worksheets similar to the model with an appropriate number of plus signs and an equal sign for each trial. (See illustration 2.)
- Show the children how to place bear counters on the circles on the worksheet. Count the number of bears to the left of the plus sign and the numbers of bears to the right of the plus sign. Count the combined sets and compare to the set on the right of the equal sign.
- Roll two dice and help the children practice counting the number of dots in the sets on the top faces of the two dice.

What Children Do
1. Add sets of bears by:
   - placing bears on the circles to the left of the plus sign and the right of the plus sign;
   - counting the bears on each side of the plus sign on the worksheet; (See illustration 1.)
   - counting the number of bears on the right of the equal sign and comparing it with the total number of bears on the left of the equal sign; and,
   - continuing the activity with new sets of bears.
2. Roll a handful of dice; Then:
   - match dice with sets of one, two, etc.; on top
   - place bear counters by each group to show the arrangement of the dots on the dice in the group; and,
   - count the number of bear counters for each group. (See illustration 2.)
3. Roll two dice; Then:
   - clap hands and count out loud, 1, 2, 3, etc., for each dot on the two dice; and,
   - repeat the activity with new rolls of the dice.

More To Do
- For more advanced children provide more dice and plus signs for each trial, so they can count to higher numbers.
- Have the more advanced children learn addition facts by doing activity 2 above and reciting the addition facts created such as "a set of 2 plus a set of 3 equals a set of 5."

More Checkers Sets

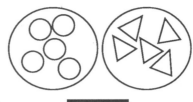

What Children Learn
- Identifying like sets despite differences in objects used, spacing or arrangement.

Before You Begin
- Place all materials on a table.

What Children Do
Respond to questions about sets.

"WHICH ROW HAS MORE?"

1. Teacher places 5 checkers on the table in a straight line so that they touch; places 5 checkers below them, but leaves space between them. Asks, *are there more checkers in the long row or the short row?* Child checks her answer by counting the checkers in each row. (Both rows have the same number of checkers, even though it looks like the long row might have more.)
2. Teacher places 10 checkers on the table in a straight line; places 10 checkers in a circle. Ask, *are there more checkers in the circle or the line?* Child checks her answer by counting the checkers in the line and in the circle. (Both arrangements have the same number of checkers, even though they are arranged differently.)
3. Repeat with different numbers and arrangements of checkers.
4. Teacher places six large items (e.g. apples) and 6 small items (e.g. poker chips) on the table and asks, *which has more?* Child checks her answer by counting the apples and poker chips. (Both have the same number of items, even though one set has larger items than the other.)

More To Do
- Work in pairs. They take turns creating two equal sets, using the items provided, and asking the other child, *which has more?*

One Little, Two Little, Three Little Blocks...

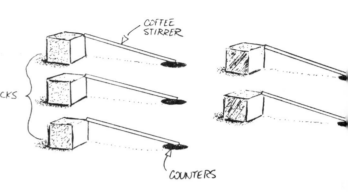

Fives

What Children Learn
- To compare sets, with up to six objects, and determine which has **more** or **less**.
- To estimate the number of objects in a given set.
- To use counters to determine whether or not the estimate is correct.

Before You Begin
- Discuss the idea of **sets**. We make sets by grouping things. We can create sets which have certain things in common. For example, we could put red blocks in one set and blue blocks in another. We could create a set which includes all blocks. That set includes the set of red and the set of blue blocks. We can compare sets by many attributes, such as color and size.
- Show children several sets of blocks with different numbers in each group. Ask, *are these groups the same or different? How?*
- Give each child several blocks. Have the children point to the blocks one-by-one as they sing (to the tune of Ten Little Indians): "One little, two little, three little dots... etc."
- Discuss the idea of **more**. Select sets of objects in the classroom and decide which has more. *How can you check?* Encourage the children to generate strategies for checking.
- Explain that they will now play a guessing game with the blocks. Introduce the word "estimate" as another word for "guess."

Materials Needed
- Blocks in different colors
- Coffee stirrers
- Counters in two colors

What Children Do
1. Have each child:
   - put blocks into two different-sized sets (groups);
   - select a set and guess how many blocks are on the set;
   - match a counter to each block in the set by placing a coffee stirrer on each block and a counter on the other end of the coffee stirrer (use counters in one color); and,
   - repeat process with the other set of blocks and a set of of counters in a different color;
2. Guess which die has **more** blocks;
   - stack the plastic counters and,
   - based upon height, check which set has **more** blocks.
3. Repeat above procedure and guess which has **less**.
4. Repeat with different sets of blocks and guesses of **more** and **less**.

More To Do
- Place between one and six pennies in several jars. Have children guess (estimate) how many pennies are in each jar. Remove pennies from the jars and have children count them. How close were the guesses?

Simon Says..."Count to Twenty"

What Children Learn
- One-to-one correspondence.
- Following instructions and game rules.

Materials Needed
- Twenty blocks for each child

Before You Begin
- Explain that children will play "Simon Says..." using numbers.
- Have children stand up, facing "Simon" as in the traditional game.

What Children Do
Respond to "Simon's" instructions with their bodies. (Do this with the children and count aloud slowly.) Examples:

> **Simon Says...**
> Clap your hands eleven times.
> Tap your foot twelve times.
> Turn around thirteen times.
> Touch your toes fourteen times.
> Touch your nose fifteen times.
> Hop on one foot sixteen times.
> Jump seventeen times.
> Blow eighteen kisses to your neighbor.
> Tap your head nineteen times.

More To Do
- Give each child twenty blocks. Repeat the game, using blocks. Examples:

> **Simon Says ...**
> Build a tower with twelve blocks
> Give thirteen blocks to your neighbor.
> Place fourteen blocks on the floor under your chair.
> Make a train with nineteen blocks.
> Build a house with twenty blocks.

My Red and Green Pictures

What Children Learn
- How to count.
- Color names.
- To use counting for non-related activities.

Before You Begin
- Select magazines with things in the above colors.
- Look at the counters with the children. Name the colors, match counters of the same color and match counters to pictures in magazines with the same color.

Materials Needed
- Green, red, yellow, and blue counters
- Magazines with colored pictures
- Scissors
- White construction paper and glue

What Children Do
1. Sort out green and red counters, then:
   - cut green and red things from magazine pictures and glue green items onto one piece of construction paper and the red items onto another piece of construction paper;
   - match a green counter to each green item cut from the magazine and a red counter to each red item cut from the magazine;
   - count the number of green things and the number of red things they have glued onto each piece of construction paper and determine the combined total; and,
   - determine the number of red counters, the number of green counters and the combined total.
2. Count all the children wearing green, red, etc.

More To Do
- Repeat the activity with other color pairs at another time.

Cookie Jar Counting

What Children Learn
- One-to-one correspondence.
- Recognizing and ordering cardinal numerals 1-20 in sequence.

Before You Begin
(Note: This is a good game to play during outdoor time in the playground.)
- Attach a large numeral to each child's shirt with a safety pin. (Twenty children at a time can participate. Give each child a different numeral every time she plays the game.)
- Line children up in sequential order in a circle. (Once children have had practice recognizing the numerals, have them "line themselves up" in sequence, without adult help, for future games.)

Materials Needed
- Numerals 1-20, each printed on a piece of newsprint, and safety pins.
- Crackers and a "cookie jar"

What Children Do
Clap hands in unison to create a rhythm as they chant:

All:	Who stole the cookie from the cookie jar?
Teacher:	Number one stole the cookie from the cookie jar.
Child #1:	Who, me?
All:	Yes, you!
Child #1:	Couldn't be!
All:	Then, who?
Child #1:	Number two stole the cookie from the cookie jar.
Child #2:	Who, me?
All:	Yes, you!

(Game continues around the circle in sequential, not random, order. Child #20 replies that #1 stole the cookie and the cycle begins again.)

More To Do
- At snack time, place crackers in a "cookie jar." Have children "sneak up" to the jar, "steal" a cracker and take it to their seats for snack. Before eating, have children count themselves and then count the crackers. *Are here the same number of children as crackers?* Refill the "cookie jar" and repeat until all children have "stolen" and counted crackers. Eat and enjoy!

Play a Board Game

What Children Learn
- To count the dots on a die or dice.
- To use dice to play a game.

Before You Begin
- Prepare simple game boards with 20-30 steps on the path.
- Group children in twos, threes and fours depending, upon the number of game boards available.
- Show children a die. Ask, *do you know what this is? Have you ever seen one before? Where? What is it used for?*
- Model rolling the dice and moving a marker the appropriate number of stops on the path.
- Allow the children to play with the dice and compare the different sides of the dice.

Materials Needed
- Game board with a path and markers
- Large dice

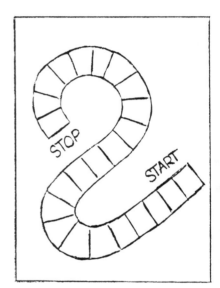

What Children Do
1. Have one child in each group:
   - roll a die to see how far to move on the board;
   - count the number of dots showing on the die;
   - move the indicated number of spaces on the board; and,
3. Take turns until they all reach the end of the path.

More To Do
- As children develop more advanced counting skills put more stops on the path and use two or three dice, so they can learn to count to higher numbers.
- Place simple board games in your math center for children to play as they practice newly-acquired counting skills (e.g. *Chutes and Ladders, First Games*).
- Use the game boards you have made to invent their own games which include rules of play, things to do when landing on a particular space (e.g. go back/ahead three spaces, miss a turn, pick a card from a stack and do what it says), etc.

Checkers Chance

What Children Learn
- Skills for making predictions (the more there is of a something in a given set, the greater the likelihood of its being chosen by random selection).

Before You Begin
- Place all checkers into the sock and mix them up.
- Give each child, and yourself, checkers to use for counting.

Materials Needed
- Several sets of checkers and a large sock

What Children Do
1. One child guesses what color checker she thinks she will get if she reaches into the sock, without looking. (Place one of your checkers of that color in front of you and say that you will keep track of the guesses.)
2. Child then selects a checker from the sock and places a checker of the same color from her stack in the center of the table. (Say that this stack will be used to keep track of checkers which are pulled from the sock.)
3. Child puts the checker she selected back into the sock.
4. Mix the checkers up and repeat until each child has had several opportunities to make guesses, select checkers and add checkers to the counting stack.
5. Look at the stacks of checkers used for counting. *Which stack is higher? Which color was picked from the sock most often?*
6. Place the stacks of checkers representing guesses next to the stacks of counting checkers. *Which is higher? How close were the guesses?*

More To Do
- Repeat the game using twice as many red checkers as black. Have children compare the stacks again before you place them into the sock. *Which has more?* Compare the results of this game to those of the previous game. *Were more red checkers picked from the sock this time? Why?*

Rock-Around-the-Clock

What Children Learn
- How to recognize and order cardinal numbers, 1-12, in sequence.
- That clocks measure the passage of time.

Before You Begin
- Make a large, paper clock on a piece of newsprint with magic marker and cut it out. Make and attach movable hands with a paper fastener.
- Print large numerals (1-12) on newsprint and cut them out.
- Attach a numeral to each child's shirt with a safety pin. (Twelve children at a time can participate. Give each child a different numeral every time he plays the game.)
- Lay the paper clock down on the floor.
- Ask questions like, *what do you know about time? What time do you go to bed? What do you know about clocks?* Encourage discussion and offer explanations, as needed, based on children's responses.

Materials Needed
- Large sheets of newsprint, a marker, scissors, safety pins and paper fasteners
- Recording of "Rock Around the Clock" by Chubby Checker
- Record or tape player

What Children Do
1. Form a circle around the clock by matching their numerals to those on the clock. *(Once children have had practice recognizing the numerals, they can "line themselves up" in clockwise sequence for future games, without using the paper clock for a cue.)*
2. Dance to the recording of "Rock Around the Clock." Each child jumps in and out of the circle as his numeral is named in the song.

More To Do
- Make paper plate clocks with crayons, attaching cut-out construction paper hands with paper fasteners.